CLIMATIC VARIATION IN EARTH HISTORY

Eric J. Barron
Earth System Science Center
Pennsylvania State University

UNIVERSITY SCIENCE BOOKS
SAUSALITO, CALIFORNIA

University Science Books
55D Gate Five Road
Sausalito, CA 94965
Fax: (415) 332-5393

Managing Editor: Lucy Warner
Editor: Louise Carroll
NCAR Graphics Team: Justin Kitsutaka, Lee Fortier, Wil Garcia,
Barbara Mericle, David McNutt, and Michael Shibao
Cover Design and Photography: Irene Imfeld
Compositor: Archetype Typography, Berkeley, California

This book is printed on acid-free paper.

Copyright © 1996 by University Corporation for Atmospheric Research. All rights reserved.

Reproduction or translation of any part of this work beyond that permitted by Section 107 or 108 of the 1976 United States Copyright Act without the permission of the copyright owner is unlawful. Requests for permission or further information should be addressed to UCAR Communications, Box 3000, Boulder, CO 80307-3000.

Library of Congress Catalog Number: 95-061063

ISBN: 0-935702-82-2

Printed in the United States of America

10 9 8 7 6 5 4 3 2

CLIMATIC VARIATION
IN
EARTH HISTORY

A Note on the Global Change Instruction Program

This series has been designed by college professors to fill an urgent need for interdisciplinary materials on the emerging science of global change. These materials are aimed at undergraduate students not majoring in science. The modular materials can be integrated into a number of existing courses —in earth sciences, biology, physics, astronomy, chemistry, meteorology, and the social sciences. They are written to capture the interest of the student who has little grounding in math and the technical aspects of science but whose intellectual curiosity is piqued by concern for the environment. The material presented here should occupy about two weeks of classroom time.

For a complete list of modules available in the Global Change Instruction Program, contact University Science Books, Sausalito, California, fax (415) 332-5393. Information about the Global Change Instruction Program is also available on the World Wide Web at http://home.ucar.edu/ucargen/education/gcmod/contents.html.

Contents

Preface ix

Climatic Variation in Earth History 1

First Case Study. Plate Tectonics and Climate: Episodes of Extensive Glaciation and Extreme Global Warmth 3

Plate Tectonics and the Surface Earth 3

How Can Plate Tectonics Influence Global Climate 6

Modeling to Explain the Extreme Warmth of the Cretaceous 9

Second Case Study. The Orbit of Earth and Pleistocene Glacial Rhythms 13

The Orbital Elements and the Insolation of Earth 15

Explanations of Glacial Rhythms 17

Summary 21

Glossary 22

Recommended Reading 23

Index 24

Preface

The geologic record contains a rich and diverse history of climatic change. This module presents two case studies from Earth history and describes only a small sample of the recorded inventory of trends and events that can contribute to our understanding of Earth as a system—as a set of interacting processes that operate over a wide range of space and time scales. The first case study examines the contrasts between major episodes of warm, apparently ice-free climates and times of major glaciation. The second case study examines the rhythms within the most recent period of glacial climate. The primary driving forces and their time scales are very different in the two case studies, but each provides substantial insight into the sensitivity of Earth to change. In both cases, the primary forcing factor pertinent to each process is insufficient to cause the "signal" recorded in the geologic record, and, therefore, we must understand a host of other factors.

Eric J. Barron, Director
Pennsylvania State University

Acknowledgments

This instructional module has been produced by the the Global Change Instruction Program of the Advanced Study Program of the National Center for Atmospheric Research, with support from the National Science Foundation. Any opinions, findings, conclusions, or recommendations expressed in this publication are those of the author and do not necessarily reflect the views of the National Science Foundation.

Executive Editors: John W. Firor, John W. Winchester

Global Change Working Group

Louise Carroll, University Corporation for Atmospheric Research
Arthur A. Few, Rice University
John W. Firor, National Center for Atmospheric Research
David W. Fulker, University Corporation for Atmospheric Research
Judith Jacobsen, University of Denver
Lee Kump, Pennsylvania State University
Edward Laws, University of Hawaii
Nancy H. Marcus, Florida State University
Barbara McDonald, National Center for Atmospheric Research
Sharon E. Nicholson, Florida State University
J. Kenneth Osmond, Florida State University
Jozef Pacyna, Norwegian Institute for Air Research
William C. Parker, Florida State University
Glenn E. Shaw, University of Alaska
John L. Streete, Rhodes College
Stanley C. Tyler, University of California, Irvine
Lucy Warner, University Corporation for Atmospheric Research
John W. Winchester, Florida State University

This project was supported, in part, by the
National Science Foundation
Opinions expressed are those of the authors
and not necessarily those of the Foundation

CLIMATIC VARIATION
IN
EARTH HISTORY

Climatic Variation in Earth History

The climate of Earth is far from constant. A rich record of climatic change is preserved in rocks and sediments, from episodes of extensive glaciation to time periods when there is little or no evidence for year-round ice. In order to understand this Earth history, we must view Earth as a system—a set of interacting processes that operate on a wide range of space and time scales (Figure 1).

Climatic variation is the product of complex interactions involving Earth's two major heat engines, the Sun and the radioactivity of Earth's interior. Solar input is the primary source of energy that drives the motion in Earth's fluid envelope, the oceans and atmosphere. The energy from Earth's interior fuels plate tectonics, leading to continental motions, volcanism, mountain building, and large-scale sea-level

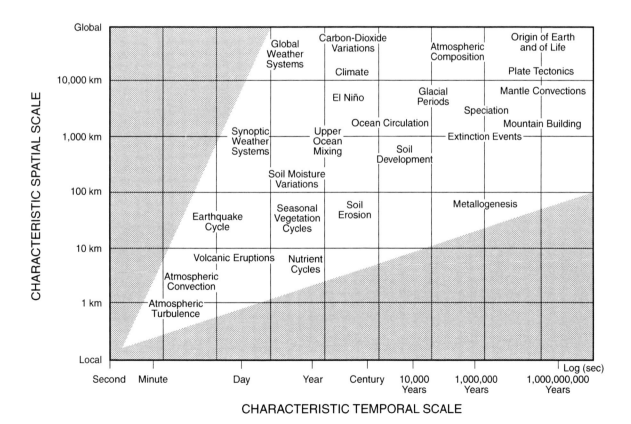

Figure 1. The characteristic time and space scales of the spectrum of Earth processes. From NASA Earth System Sciences Committee, Earth System Science: A Closer View, *1988.*

variations. The interactions between the solid Earth and the fluid Earth are reflected in the surface features of Earth and define the reservoirs and fluxes of biogeochemical cycles. The climatic record appears best explained by a combination of more than one of three basic factors: (1) the amount and distribution of solar energy received at the top of the atmosphere, (2) the composition of the atmosphere, and (3) the nature of the surface of Earth.

Two "case studies" of climate and climatic change during Earth history illustrate this view of the Earth system. One involves the major episodes of glaciation and extreme warmth, and the second examines the glacial rhythms within the current episode of glaciation. Interestingly, although the primary driving forces and their time scales are quite different, atmospheric carbon dioxide appears to play a role in each of the case studies.

As the two examples will illustrate, continued study of Earth's rich record of climatic change and Earth system interactions promises to provide many new insights into global evolution and the sensitivity of Earth to change.

FIRST CASE STUDY
Plate Tectonics and Climate— Episodes of Extensive Glaciation and Extreme Global Warmth

The most persuasive evidence for global climatic change in Earth history is the record of extensive glaciations separated by periods for which there is little or no evidence of year-round ice (Figure 2). From an Earth-history perspective, the climate of today is distinctly glacial. Much of the research that attempts to explain the major changes in climate, illustrated in Figure 2, focuses on periods of the greatest contrast. The warm climate of the Cretaceous period (approximately 100 million years ago) exhibits the largest well-documented contrast to the present glacial climate.

The evidence for global warmth comes from every facet of the geologic record including paleontology, geochemistry, and sedimentology. More than 400 plant species are recorded from latitudes above the Arctic Circle, and these polar floras are indicative of seasonal conditions with mean annual temperatures between 5 and 10°C. Fossils of large ectotherms (cold-blooded organisms) are found at latitudes as high as 60 degrees, whereas we know that modern relatives (alligators and crocodiles) become inactive below 20°C and are restricted to much lower latitudes. In the marine realm, no evidence of cold-water faunas from Cretaceous time has been discovered. Paleotemperature data suggest that deep ocean temperatures were between 15 and 17°C, compared to modern values of 1–2°C. Tropical surface temperatures were evidently similar to modern values, or a few degrees higher. A combination of these data (Figure 3) suggests that the globally averaged surface temperatures were 6–12°C higher than at present, with polar temperatures 20–50°C higher.

Not only did the Cretaceous show substantially greater warmth over millions of years, but the transition from the warm climatic conditions of the Cretaceous to the modern glacial climate occurred slowly, over tens of millions of years. This evidence argues for a causative mechanism that operates over a long time. Since the formulation of the plate tectonic theory of crustal evolution, the changing distribution of continents has become a frequently cited explanation for the occurrence of glacial and nonglacial climates. The following sections explain how.

Plate Tectonics and the Surface Earth

Earth's crust (lithosphere) consists of a series of rigid plates that are in extremely slow but constant motion. Plate tectonics is defined simply as the interactions between these plates. Three types of interactions occur: (1) two plates diverge, and new lithospheric crust is created from hot molten rock that moves up from below; (2) plates converge, forcing one beneath the other in a process known as subduction; and (3) two plates slide past each other without

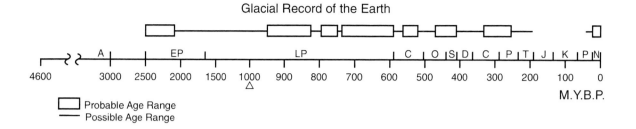

Figure 2, above and right. The probable time range of glacial episodes during Earth history. Times of uncertainty in interpreting the record are given as possible ranges of glaciation.

Geologic Time Scale

Era	Period	Epoch	Years Before the Present
Cenozoic	Quaternary	Holocene (Recent)	
			11,000
		Pleistocene (Glacial)	
			1,600,000
	Tertiary	Pliocene	5,300,000
		Miocene	23,700,000
		Oligocene	36,600,000
		Eocene	57,800,000
		Paleocene	
			66,400,000
Mesozoic	Cretaceous		144,000,000
	Jurassic		208,000,000
	Triassic		245,000,000
Paleozoic	Permian		286,000,000
	Carboniferous — Pennsylvanian (Upper Carboniferous)		320,000,000
	Carboniferous — Mississippian (Lower Carboniferous)		360,000,000
	Devonian		408,000,000
	Silurian		438,000,000
	Ordovician		505,000,000
	Cambrian		570,000,000
Precambrian	Proterozoic		2,500,000,000
	Archean		3,800,000,000

Figure 3. Mean annual surface temperatures in Cretaceous times, compared with modern values (after Barron, 1983). Some of the major constraints are based on oxygen isotopes (in benthic and planktonic Foraminifera, bottom-dwelling and floating microorganisms that secrete a $CaCO_3$ shell, and belemnites, ancient relatives of a nautilus), reef distribution, and the absence of year-round ice.

convergence or divergence. The idealized plate in Figure 4 includes both continental and oceanic crust. The plate is growing with the addition of new material (accretion) at the midoceanic ridge on the right, and oceanic crust is being destroyed (subducted) where the idealized plate converges with the adjacent plate on the far left. Throughout the world, subduction is indicated by the distribution of deep-sea trenches, by explosive volcanism, and by deep-seated earthquakes. In Figure 4, the ocean basin to the right of the continent is growing in area while the ocean basin to the left of the continent decreases in size.

One consequence of plate tectonics is that the distribution of continents is continually changing. The Atlantic Ocean first formed about 165 million years ago, at which time the world consisted of a single megacontinent, Pangaea, and a single super-ocean called Panthalassa. We can find times in Earth history when Greenland straddled the equator and times when the Sahara Desert was at the South Pole.

The topography or elevation of the continents is also governed by plate tectonics. With continued motion of the plates, collision between continents is inevitable. Usually, the thinner and more dense oceanic lithosphere is subducted beneath the thicker and less dense continental lithosphere. For example, the existence of the Andes mountain range is a direct consequence of South America's being at the boundary of an active or converging plate. The Himalayan-Tibetan Plateau region, the highest topography in the world, is the product of the collision of India and Asia nearly 50 million years ago.

Plate tectonics also influences sea level. At boundaries where new crust is created (predominately at midocean ridges), hot, thick lithosphere is accreted to the margins of diverging plates. Over tens of millions of years the thick lithosphere cools and contracts. In fact, the new crust is clearly evident as ridges (midocean ridges) on maps of the depth of the ocean. Consequently, at times when new sea floor is created more rapidly by faster spreading at divergent boundaries or increased length of midocean ridges, the holding capacity of the ocean basins decreases and some continental flooding occurs. For example, most estimates suggest that during the middle Cretaceous new sea floor was created at a rate twice as fast as it is today. The theory of lithospheric cooling and contraction suggests that such a fast sea-floor spreading rate would yield a global sea-level rise of 200 to 300 meters (this may be compared with 65 meters if we melted the modern ice caps). In fact, almost 20% of the modern continents was covered by ocean in the middle Cretaceous (Figure 5). In short, plate tectonic processes affect not only the distribution and elevation of the continents, but also the area of continents above sea level.

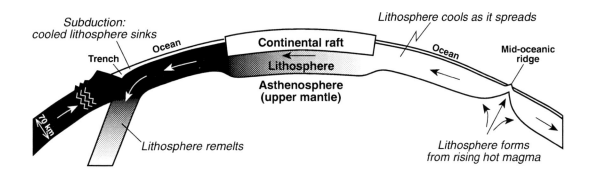

Figure 4. From right to left, the motion of lithospheric plates in response to mantle convection.

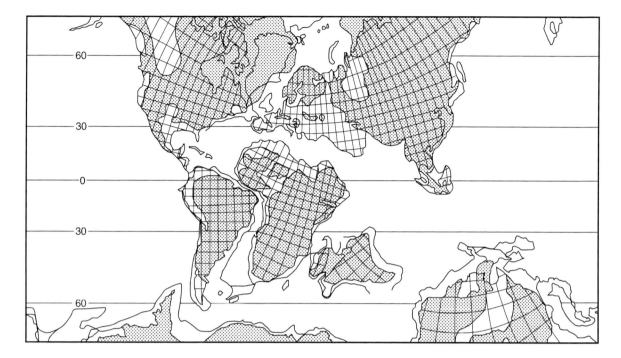

Figure 5. A reconstruction of mid-Cretaceous geography extending to 70° north and south latitudes (Barron et al., 1981). Present continental outlines and the 2,000-m depth contour are included with mid-Cretaceous continental area (shaded). Paleolatitude lines are given at 30° intervals, and the continents include their present-day latitude/longitude lines for reference.

How Can Plate Tectonics Influence Global Climate?

The distribution, elevation, and areal extent of the continents each influence climate. In addition, plate tectonics influences climate by modifying the atmospheric concentration of carbon dioxide.

Continental Distribution

In assessing the effects of the distribution of continents on global climate, five major factors must be addressed: differences in surface albedo, land area at high latitudes, the transfer of latent heat, restrictions on ocean currents, and the thermal inertia of continents and oceans. Each of these factors is explained below.

Albedo. Different types of surfaces reflect different amounts of incoming solar radiation; for example, a snow-covered field reflects more sunlight than a dark soil. The amount of energy that a surface reflects is called its albedo. The albedo of a land surface is very different from that of an ocean surface. The average albedo of a tropical ocean is near .07 (7% reflected, 93% absorbed), while that of a desert region may approach .25. Considering that the amount of incoming solar energy is very different at different latitudes (it is greatest at the equator), the latitudinal distribution of land and sea has a strong potential to influence the total energy budget of Earth.

High-latitude land area. Land at high latitudes provides a surface for the accumulation of snow and year-round ice. Snow and ice have a very high albedo; for fresh snow it approaches .65 to .80. Consequently, high-latitude land area, if snow-covered, can also dramatically influence the energy budget of the atmosphere-ocean system.

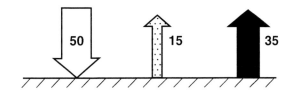

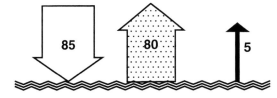

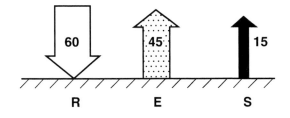

Figure 6. Heat balances for some representative surface types. On average, net radiative heating of the surface is balanced by evaporative and sensible heat loss. Evaporation is the preferred mode of heat loss at warm temperatures; thus subtropical oceans lose heat mainly through evaporation.

Transfer of latent heat. Changes in albedo are not the only mechanism by which continental distribution affects the global energy budget and, hence, the global climate. An examination of the surface energy budget, in terms of sensible and latent heat fluxes (Figure 6), reveals a third means by which continental distribution influences climate. The largest energy fluxes at the surface of Earth involve moisture (latent heat flux). Clearly, the distribution of land and sea will influence evaporation and precipitation and therefore the total energy budget of the atmosphere.

Restrictions on ocean currents. Continents act as barriers to the flow of oceanic currents. The oceans transport a substantial amount of heat poleward (a maximum of $2-4 \times 10^{15}$ watts at latitudes of 20–30 degrees, representing one-third to one-half of the total poleward heat transport by the oceans and the atmosphere). The distribution of land can block poleward heat transport by the ocean and may influence polar climates and the subsequent extent of ice and snow cover. Thus, the shape and size of the ocean basins become a factor in controlling global climate.

Thermal inertia. A continental surface has little thermal inertia. Basically, a continental surface responds rapidly to the current solar input. In contrast, the oceans have a large heat capacity. They store solar energy in summer and release it in winter. Therefore, the thermal inertia of the oceans tends to moderate the role of the seasonal cycle of insolation. The range of summer or winter temperatures could influence whether or not snow accumulated in the winter or melted in the summer. Again, in this example, the distribution of land and sea may be a controlling factor in the occurrence of glacial or nonglacial climates.

Continental Elevation

Climate is also influenced in a number of ways by the elevation of continents. Examples of this factor include temperature change and varying moisture content with changing elevation, the effect of mountains on large-scale atmospheric circulation, the effect of mountains on regional climate, and influences of topography on the general circulation of the atmosphere.

Effects of changing elevation. Atmospheric temperature decreases with height, a decrease referred to as the lapse rate. The average lapse rate is 6.5°C per kilometer of height. The colder temperatures at higher elevations promote snow accumulation, and therefore mountains become a base region for glaciation. Again, snow and ice have a very high albedo, so the occurrence of year-round snow and ice promotes cooling through modification of the energy balance.

Effects on large-scale circulation. The distribution of mountains influences the large-scale circulation of the atmosphere. In the simplest form, this influence is analogous to that of a large rock in a stream. The rock acts as a barrier to the flow of the fluid, and the current pattern is modified around and downstream from the barrier. The positions of the continents and oceans (causing differences in heating) and the distribution of regions of high topography control the position of the large-scale waves in the atmosphere, such as the jet stream, and therefore control the pattern of the weather. A change in topography may well control the distribution of cold air masses or the track of winter storms. Such changes could initiate glaciation in a particular region by promoting even greater cooling, or they could warm high-latitude regions which may otherwise be cool.

Effects on regional climate. The regional differences in climate associated with topography might be considered a third mechanism of climatic change. The windward sides of mountains tend to be much wetter, while a rain-shadow effect occurs in the lee of mountains. On the windward side the air is forced upward, cooling with increasing elevation. Since cooler air can hold less moisture, the cooling air mass reaches saturation and precipitation occurs. In the lee, deserts develop because the sinking air warms and therefore the level of saturation decreases as the air descends. For this reason, the different sides of mountain belts tend to have very different climates.

Effects on general circulation. Topography can exert an influence on the general circulation of the atmosphere or large-scale wind patterns. The general circulation is very complex, but it can be understood by a series of thought experiments. First, imagine a nonrotating Earth. On average, the air near the equator is heated the most and the air near the poles the least. In response to this, the cold, dense air at the poles sinks and spreads toward the equator, while the air over the equator rises and moves toward the poles. In this case the Northern Hemisphere winds near the surface would be from the north and the winds at higher altitudes would be from the south.

What happens if Earth rotates about its axis? Every point on Earth makes a complete circuit in a day. Near the poles, each point on the Earth's surface hardly moves; points on the equator, being farthest from Earth's axis, move most rapidly. What does this mean for the wind? Consider the case of air moving northward from the equator. The surface north of the equator moves more slowly, so the air moves toward the east—the winds are now from the southwest instead of the south. Similarly, the sinking cold air at the poles has almost no eastward speed, but south of the pole, the surface beneath moves eastward faster. Thus the polar air, which started out moving southward, begins to turn southwestward. The winds are from a northeasterly direction.

Because of the rotation of Earth, the circulation in each hemisphere tends to be broken up into three cells. On the surface of the Northern Hemisphere, the winds near the poles are from the northeast (the polar easterlies), and the winds near the equator are from the northeast (the northeast trade winds). They are separated by a broad zone of west winds in the middle latitudes (the midlatitude westerlies). The easterlies and westerlies exert a frictional drag on the surface of the earth that must balance.

Now let's add mountain belts. A mountain belt in the easterly wind belt would exert a westward frictional drag on the winds. The opposite would occur in the region of the westerlies. Differences in pressure across mountains would play a role similar to the frictional drag. Therefore, the strength and the distribution of the easterlies and westerlies would change for an Earth with fewer mountains or a different distribution of mountains.

Carbon Dioxide and Plate Tectonics

There are numerous mechanisms by which aspects of plate tectonics directly modify the atmospheric or oceanic circulation, or directly influence the surface energy balance (albedo changes). Plate tectonic processes also modify the carbonate-silicate geochemical cycle.

Increased sea-floor spreading rates and increased plate destruction by subduction (faster plate tectonic processes) should result in greater outgassing of carbon dioxide from volcanoes. On long time scales, carbon dioxide production is balanced by consumption during weathering of igneous and metamorphic rocks. To maintain the atmospheric carbon dioxide concentration, the amount of carbon dioxide consumed by silicate weathering would also have to increase. Many researchers suggest that during the Cretaceous the rate of sea-floor spreading was nearly twice as great as the modern rate.

In addition, as stated earlier, higher rates of plate accretion are associated with higher sea level. For those times for which much higher sea-floor spreading rates are hypothesized, increased continental flooding should occur (e.g., covering 20% of the present-day continental area in the middle Cretaceous). Consequently, not only was it likely that volcanic outgassing of carbon dioxide was higher than today, but the area of continental rocks exposed that would consume atmospheric carbon dioxide through the weathering of silicate rocks was substantially less.

This simple argument suggests that changes in the rates of plate tectonic processes can produce changes in the atmospheric concentration of the important greenhouse gas carbon dioxide. Such changes would occur on a time scale of tens of millions of years. Interestingly, the times of high sea level are often associated with planetary warmth and the times of low sea level are associated with cooler climates. This association may reflect the fact that atmospheric carbon dioxide concentration is tied to plate tectonics as part of the carbonate-silicate cycle.

Modeling to Explain the Extreme Warmth of the Cretaceous

A large number of mechanisms related to plate tectonics may be responsible for, or contribute to, the occurrence of major glaciations or extreme climatic warmth in Earth history. Climate models provide an opportunity to probe whether any specific forcing factor has been important in regulating past climates. Models are mathematical representations of the climate system. By changing a specific factor, such as land-sea distribution, a scientist can examine the model response and compare it with the information derived from the geologic record.

Evaluation of all the mechanisms described above is not straightforward. In fact, multiple mechanisms may be responsible for changes in climate, and there is no reason to suppose that the same mechanism produced each of the major climatic episodes illustrated in Figure 2. Since the changes in Cretaceous geography are large and well known (Figure 5), study of the effects of the distribution of land and sea is a logical starting point to try to explain Cretaceous warmth. In order to assess the importance of geography, the model results can be compared with the surface temperature data plotted in Figure 3.

Paleogeography and Cretaceous Warmth

A hierarchy of climate models, from simple two-dimensional models that consider only the equator-to-pole temperature contrast to three-dimensional models that simulate the atmospheric circulation, has been applied to the study of the extreme climatic warmth of the Cretaceous. The results from each type of model simulation suggest that Cretaceous land-sea distribution is an important factor contributing to the extreme warmth. For example, one of the most comprehensive models of the atmosphere has been utilized to simulate the climate for the Cretaceous, based on the geography illustrated in Figure 5. The global average temperature for a simulated Cretaceous climate, based on this geography and without polar ice caps, was 4.8°C higher than for a present-day control simulation. The warming was accentuated at high latitudes (Figure 7 shows a 15°C and 35°C increase at high northern and southern latitudes, respectively), corresponding very closely to the sense of the warming estimated from the geologic record. However, this result is problematic. Even though the warming is in the right sense, the magnitude is only half of what is required to explain the geologic evidence.

Interpreting the result is more difficult because the simulated Cretaceous warming includes changes in several different model parameters at once (continental positions, topography, sea level, and removal of the ice caps), relative to the present-day control simulation. What factors are responsible for the 4.8°C warming? Modelers attempt to dissect such complex cause-and-effect relationships by repeating the experiments, changing only one variable at a time to determine which were most responsible for the net change.

Interestingly, as Figure 8 shows, the warming of the Southern Hemisphere was due almost entirely to the removal of the Antarctic ice cap. (Remember that the difference between warm, ice-free climates and glacial climates is what we are trying to explain, yet the total simulated temperature difference at high southern latitudes seems to be associated with the removal of ice rather than some geographic factor!) Most of the Northern Hemisphere warming can be attributed to the difference in the amount of high-latitude land. The model showed no sensitivity to sea level, and although topography played an important role in governing regional climates, removal of all modern-day topography did not result in a substantial change in the global temperature.

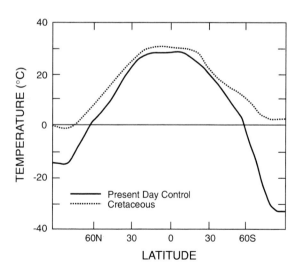

Figure 7. A comparison of the zonally averaged surface temperature (averaged across latitudinal belts) for the present-day control and "realistic" Cretaceous simulation.

The results of the climate model experiments are of considerable interest, and there are several possible ways to interpret them. First, assume that the model simulations are accurate. Clearly, then, the actual geographic changes result in only a modest Northern Hemisphere warming and the cause of the difference between global warmth and glaciation remains unsolved. Some additional mechanism would be required to explain the extreme warmth of the Cretaceous. Alternatively, assume that the geologic data is in substantial error. Certainly many difficulties are associated with determining the temperatures prior to the historical record. However, the model-simulated warming deviates considerably from the observations; the simulations are well below even the lower limit of the geologic interpretations.

Second, assume that the modern climate models that are used here to simulate the Cretaceous climate may be missing some fundamental process or feedback mechanism, and therefore are unable to correctly simulate climates very different from today's. This is a very real possibility as mathematical models are highly idealized representations of the climate. For example, the role of the oceans and the characteristics of the ice caps are poorly represented in many climate models. Some researchers have suggested that the role of the oceans, with their tremendous ability to store heat, is a key to the problem. Because of this storage of heat, oceanic regions have a smaller seasonal change in temperature than land areas at the same latitude. Using an energy balance climate model, Thomas Crowley and his colleagues demonstrated that the presence of oceanic regions near the poles can reduce summer temperatures over adjacent polar continental regions in comparison with situations of high polar continentality (see their article in Recommended Additional Reading). These researchers suggested that cooler summer temperatures may not melt winter snow and ice, thus initiating glaciation. The results suggest that the role of geography has not yet been fully explored and that the models must link the oceans, the atmosphere, and the ice caps in order to provide an understanding of the history of Earth's climate.

What is the solution? Is the geologic record misinterpreted? Are models far too inaccurate? Do we need some additional forcing factor? The geological record and the model simulations provide one additional clue that helps to focus the investigation. Remember that global warmth appears to be well correlated with times of high sea level, yet the climate model results described earlier exhibited almost no response in an experiment that flooded 20%

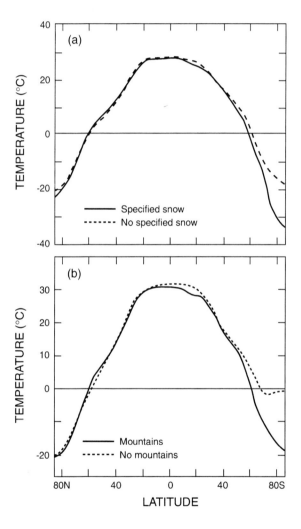

Figure 8. (a) A comparison of the zonally averaged surface temperature for the present-day simulation with and without specified permanent snow (fixed high albedo) on Antarctica and Greenland. (b) A comparison of the zonally averaged surface temperature for the present-day simulation with mountains and without mountains.

of the area of the continents. This point suggests that something fundamental is missing from our simulation. We then remember that times of continental flooding are associated with rapid sea-floor spreading rates, and presumably higher carbon dioxide degassing rates.

Carbon Dioxide and Cretaceous Warmth

Although the geochemical models of the carbonate-silicate cycle indicate that changes in atmospheric carbon dioxide levels are plausible and that higher levels in the Cretaceous are likely, there are no accurate measurements of atmospheric gas concentrations for periods tens of millions of years ago. However, continued experimentation with the climate model suggests that quadrupling the atmospheric carbon dioxide level in the Cretaceous, in addition to the differences in geography, could produce an 8°C increase in temperature compared to the present. For the first time the model-simulated temperatures (Figure 9) are close to the limits of the observations from the geologic record. A quadrupled level of carbon dioxide for the Cretaceous is well within the limits suggested by the geochemical models of the carbonate-silicate cycle.

Of course, improved models may give very different results. The carbon dioxide level required to achieve the extreme climatic warmth of the Cretaceous may be greater than or less than four times the present level. More importantly, the experiments suggest that several factors are required to explain the climate of a single time period, and, significantly, the mechanisms are interrelated. Earth's internal heat engine, which drives plate tectonics, alters climate directly through interactions of the atmosphere and oceans with the surface of Earth, and indirectly through geochemical interactions that alter the greenhouse character of the atmosphere. Both the direct effects of plate tectonics and the influence of plate tectonics on geochemical processes are required, in concert, to explain the large-scale changes in climate—from episodes of extreme warmth to the major periods of polar glaciation.

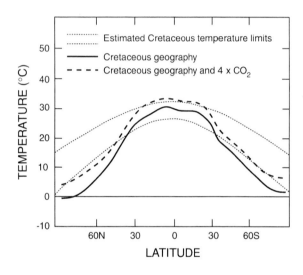

Figure 9. A comparison of the estimates of Cretaceous temperature limits (dotted lines) with predicted zonally averaged surface temperatures for Cretaceous geography (solid line) and for Cretaceous geography plus quadrupled carbon dioxide (dashed line). Predicted temperatures were derived from a general circulation model of the atmosphere (Barron and Washington, 1984).

SECOND CASE STUDY
The Orbit of Earth and Pleistocene Glacial Rhythms

The first unequivocal evidence of a large Antarctic ice cap is the occurrence of ice-rafted detritus in deep-sea cores dated at 28 million years of age. (Ice-rafted detritus is an indicator that continental ice carrying material eroded from the continents was transported out to sea by icebergs.) Greenland was clearly glaciated 3 million years ago. A closer examination of the record of glaciation during the last 1.9 million years (the Pleistocene) reveals considerable climatic variability in the form of repeated advances and retreats of the polar ice caps.

The movement of ice over the land surface is an extremely destructive process, mechanically grinding the rock surface and carrying the glacial debris downstream. On land the history of glaciation is recorded by glacial tracks and scours and by the deposition of glacial debris at the front (terminal moraines) and sides of glaciers and ice caps. In North America and Europe four major groups of terminal or end moraines are preserved from the Pleistocene. Does this record suggest four major glacial advances of the modern ice cap? Moraines give scientists only the evidence of the farthest extent of glaciation. A more recent glaciation would destroy the record of any previous ice age where the maximum extent of the ice was less than that of the most recent advance. Until 1950, the end moraines were the only record of the ice age, and only four major glaciations were identified.

A much better record was derived from the deep sea with the invention of coring devices in 1950. The coring of oceanic sediments retrieved a continuous record of microscopic fossil shells composed of calcium carbonate ($CaCO_3$). This new-found record was examined in detail with an innovative chemical methodology based on isotopes.

A chemical element is defined by the number of protons in its nucleus. The number of neutrons in the nucleus can vary. Isotopes are elements with different mass numbers defined by the total number of neutrons and protons. Because isotopes have different masses, different physical and chemical processes may separate, or fractionate, them. For example, oxygen has three isotopes: oxygen-16, -17, and -18. In the process of evaporation from the surface of the ocean, water containing oxygen with the lightest isotope value is preferentially evaporated. Likewise, in precipitation there is a preference to rain out water with the heaviest isotope, oxygen-18, first. The atmosphere is, in a sense, distilling the oxygen isotopic composition of the moisture it transports through the processes of evaporation and precipitation. The moisture that is finally deposited on the ice caps is extremely enriched in the light isotope, and therefore, during times of extensive glaciation, the ocean becomes depleted in oxygen-16. The microscopic organisms in the ocean that grow $CaCO_3$ shells preserve the oxygen isotopic composition of the oceanic water in which they live and build their shells.

The record of oxygen isotopes for the last 700,000 years (Figure 10) reveals a remarkably rhythmic record of repeated glaciation. More oxygen-18 recorded by shells indicates more of the lighter oxygen-16 stored as ice on land. Note that the oxygen-18 record for each ice age has a similar amplitude and that the record for each ice age has a ramp-like structure suggesting slow buildup to full glaciation and then rapid retreat. The current climate is best described as "interglacial" and the last major ice advance ("glacial") was approximately 18,000 years ago.

A wide variety of theories has been presented to explain these glacial cycles, involving Earth orbital variability, volcanism, variations in the magnetic field of Earth, solar variability, interstellar dust, internal oscillations, Antarctic ice surges, atmospheric carbon dioxide concentration, and deep ocean circulation changes. The remarkably rhythmic nature of the glacial cycles, undiscovered until more complete records from the deep sea were collected, has proven to be a major test for the theories of glaciation. A deterministic view of the climatic system would suggest that the glacial-interglacial rhythms were the product of some periodic climatic forcing factor.

In the first half of the twentieth century, Milutin Milankovitch, a Yugoslavian astronomer, proposed that periodic changes in Earth's orbital characteristics were the major control over glaciation. Milankovitch computed periods for the variation in the time of year during which Earth is closest to the Sun, governed by the orbital precession of the equinoxes (approximately 19,000 to 23,000 years—see discussion below); the axial tilt of Earth approximately 41,000 years); and the eccentricity of Earth's orbit around the Sun (approximately 100,000 years). Milankovitch's hypothesis was rejected at the time due to the weight of evidence from continental regions suggesting that there were only four major glaciations.

James Hays, John Imbrie, and Nick Shackleton revitalized the Milankovitch hypothesis when they calculated the timing of variations for three geologic factors—surface temperature, oxygen isotopes, and the abundance of a particular plankton species—from high-resolution, continuous deep-sea cores representing the last 700,000 years (see their paper in the Recommended Additional Reading). They found that these variables illustrated a rhythm with intervals of 19,000, 24,000, 43,000, and 100,000 years (Figure 11). The dominant climatic periodicity is 100,000 years. The correspondence of the periods from the geologic data with the periods of the Earth's orbital characteristics is unmistakable.

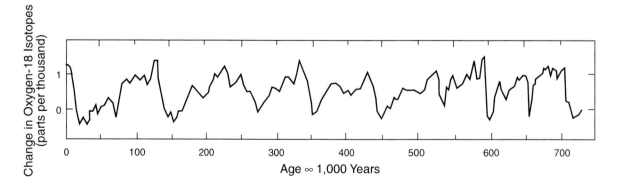

Figure 10. Oxygen isotopic record for the last 700,000 years, illustrating glacial/interglacial cycles (after Emiliani, 1978). Higher oxygen-18 levels show interglacial periods.

The Orbital Elements and the Insolation of Earth

The gravitational effects of the planets within our solar system perturb the orbit of Earth. Since the input of solar energy is dependent on the distance between Earth and Sun and on the solar angle, these orbital changes alter the seasonal and latitudinal distribution of incoming solar energy (called insolation).

Eccentricity

Due largely to interactions with Jupiter and Saturn, the orbit of Earth about the Sun varies from a circle (an eccentricity of 0) to an ellipse (a maximum eccentricity of .07), with a period of 96,600 years (Figure 11a). The Sun is currently nearest Earth in January and farthest away in July. During periods of maximum eccentricity, the differences in Earth-Sun distance from summer to winter can greatly accentuate the contrast between the seasons. However, on an annual average, variations in the eccentricity result in less than a 1% range in the total insolation.

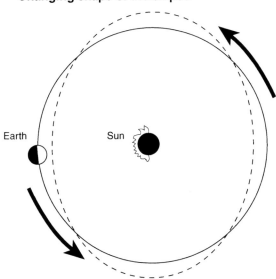

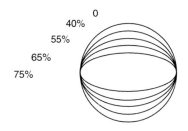

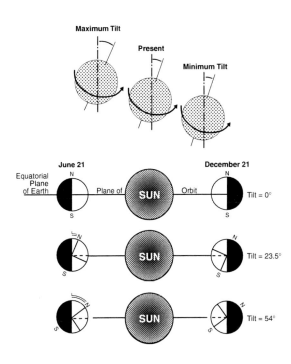

Figure 11a. The orbit of Earth changes shape from nearly circular to more elliptical. This is termed eccentricity and is expressed as a percentage (after Imbrie and Imbrie, 1979).

Figure 11b. Variation in the tilt of Earth's axis from maximum to minimum, and the effect of axial tilt on the distribution of sunlight. When the tilt is decreased from its present value of 23.5°, the polar regions, in summer, receive less sunlight; when the tilt is increased, polar regions receive more sunlight (after Imbrie and Imbrie, 1979).

Obliquity

The obliquity of Earth is measured as the tilt of Earth's axis from the normal to the plane of its orbit (Figure 11b). Today the obliquity is 23.4 degrees (the latitude of the tropics of Cancer and Capricorn). With a period of 41,000 years the obliquity varies from 22 degrees to 24.5 degrees. The condition of higher obliquity (greater tilt toward and away from the Sun) produces more pronounced differences between winter and summer seasons and greater total solar energy at the polar region. The condition of low obliquity (less tilt) reduces the seasonal cycle, but a smaller amount of insolation is received in polar regions during the year. The variation in obliquity produces a difference of approximately 10% in polar insolation.

Precession of the Equinoxes

Solar and lunar torques on the equatorial bulge of Earth cause the time of perihelion (the time at which Earth is closest to the Sun) to vary

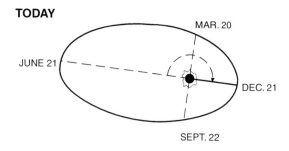

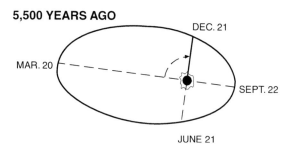

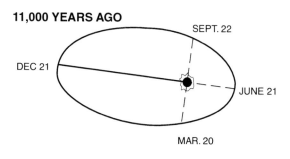

Figure 11c: Precession of the equinoxes. Owing to axial precession and to other astronomical movements, the positions of equinox (March 20 and September 22) and solstice (June 21 and December 21) shift slowly around Earth's elliptical orbit, and complete one full cycle about every 22,000 years. Eleven thousand years ago, the winter solstice occurred near one end of the orbit. Today, the winter solstice occurs near the opposite end of the orbit.

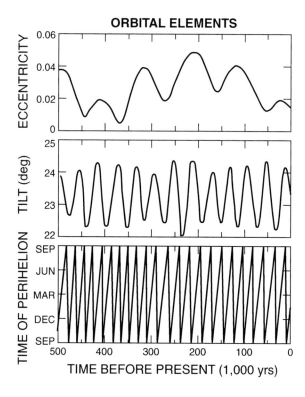

Figure 12. Variations in Earth's orbital elements, eccentricity, tilt (obliquity), and time of perihelion (precession of the equinoxes) computed for the last 500,000 years with a computer program written by Tamara Ledley and Starley Thompson following Berger (1988).

(Figure 11c). Currently, perihelion is in January, tending to moderate Northern Hemisphere winters. Ten thousand years ago, the time of perihelion was in Northern Hemisphere summer, resulting in a much more pronounced seasonal cycle. The precession of the equinoxes has a period of approximately 21,000 years. Combined with eccentricity, differences in the time of perihelion can result in differences of as much as 33 days in the length of astronomical summer and winter.

Figure 12 illustrates the variations in Earth's orbital elements—eccentricity, tilt (obliquity), and time of perihelion (precession of the equinoxes)—as computed for the last 500,000 years by Andre Berger. From the orbital variation (knowledge of the distance between Earth and Sun and the solar angle) the variations in insolation (Figure 13) can be determined. There is little change in the global, annual-average insolation; however, there are relatively large changes in the amplitude of the seasonal cycle and in the receipt of solar energy at high latitudes. In Figure 13 the Northern Hemisphere insolation varies by as much as 9% on either side of the mean insolation value.

These calculations can be extended back in time for approximately 5 million years, at which point phase relationships and the amplitude of the variations cannot be accurately reconstructed.

Explanations of Glacial Rhythms

The acceptance of Milankovitch's hypothesis, that orbital cycles provided the pacemaker for the ice ages, provides the basis for a unique case study of global climatic change. The mechanism is apparently known and the geologic record illustrates a well-defined "response," which can be the basis for examining climatic sensitivity to an external forcing factor. Much of the research has focused on developing the physical explanations required to substantiate the statistical association demonstrated by Hays, Imbrie, and

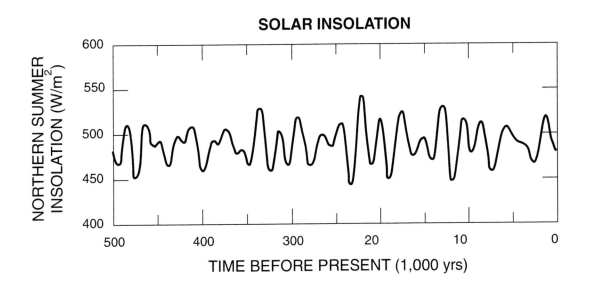

Figure 13. Variations in insolation (in watts per square meter) determined from the variations in Earth's orbital elements.

Shackleton in 1976 (see their paper in the Recommended Additional Reading).

The complex general circulation models (GCMs) utilized in the investigation of Cretaceous warmth require too much computer time to simulate the thousands of years of glacial climate. Climatologists instead have turned to simpler models governed by consideration of Earth's energy balance. Max Suarez (University of California, Los Angeles) and Isaac Held (NOAA Geophysical Fluid Dynamics Laboratory) made one of the early attempts to verify the Milankovitch theory with a climate model. Their energy balance climate model was zonally symmetric and included seasonal changes in the insolation.

In this model, changes in snow cover (simply based on predicted surface temperature) and associated changes in albedo play the dominant role in determining the sensitivity of the model to orbital changes. The model results showed substantial sensitivity to orbital variations (a 2.4°C difference in Northern Hemisphere temperature between different orbital conditions). The results indicated significant sensitivity to obliquity and precession.

However, two problems were evident: (1) the models failed to reproduce the dominant 100,000-year cycle of glaciation and the ramp-like nature of the glacial growth and decay, and (2) the models failed to reproduce a lag between insolation changes and the buildup and decay of glaciation. The results are not unexpected. Eccentricity alone does not have a major influence on insolation, and a 100,000-year rhythm is not evident in the insolation calculation illustrated in Figure 13. What could explain the discrepancies?

The discrepancies noted from the early attempts to simulate glacial rhythms are a road map to improvement of climate models. The simulation's lack of a lag between cause and effect suggests that some of the long time constants of the climate system must be considered, such as deep-ocean processes and ice-sheet dynamics. The inability to simulate the 100,000-year signal guides us to a consideration of other forcing factors and other amplifiers or modulators of the climatic response. In summary, the research on the ice ages began to focus our attention on some of the slower "physics" of the climatic system, most notably ice-sheet dynamics.

The first step was to add an ice sheet to the energy balance model in order to include climatic system inertia based on the volume of the ice sheets. When the model was run, the simulated changes in ice-sheet volume were 20 to 50% of the observations (Figure 14), and only the oscillations due to obliquity and precession were apparent. The second major step was to include the response of the lithosphere to ice-sheet buildup. Essentially, the weight of the ice sheet depresses the lithosphere.

Johannes Oerlemans of the University of Utrecht, Netherlands, suggested that after extensive growth of the ice sheet, the slow sinking of the bedrock might bring much of the ice sheet below the snowline, resulting in rapid deglaciation. When they set the bedrock response time to ice loading at 10,000 years, they were able to simulate the ramp-like nature of the 100,000-year glacial cycle.

Problems still remain, however. Although the simulations are now qualitatively in line with the observations, they do not achieve the necessary magnitude of response to orbital variations. The most logical explanation is that the models are far too simple. Perhaps such factors as carbon dioxide concentration, topography, snow accumulation, or seemingly minor effects such as how fast ice sheets "calve," or break apart, must be included before the simulations provide a full explanation.

The association of glacial rhythms with variations in atmospheric carbon dioxide concentrations is particularly fascinating. The ice sheets consist of layers representing thousands of years of annual accumulation. Air bubbles trapped within the ice record the composition of the ancient air. The composition of the air throughout glacial cycles has been measured from ice cores, revealing a 200- to 280-ppm variation in carbon dioxide from glacial to interglacial conditions (Figure 15). Lower atmospheric carbon dioxide concentration during glacial episodes should intensify the response to orbital variations. Many hypotheses have been offered to explain these variations, including changes in plant productivity, changes in the deep-ocean circulation, and variations in nutrient availability.

Ice cap–bedrock feedbacks and the effect of atmospheric carbon dioxide concentration are not the only mechanisms under investigation to explain glacial cycles, nor is the scientific community in full agreement on the role of Milankovitch orbital variations. For example, several researchers have shown that different feedbacks, if sufficiently out of phase, can yield a limited cyclic behavior. In short, the complex interactions of the climatic system may produce rhythms even though the causative factors were not periodic.

Importantly, the glacial rhythm "case study" of climatic change suggests the interrelationships in the climatic system. Knowledge of several factors, in addition to the direct effect of orbital variations, is required to understand the rhythm of the ice ages, including ice-sheet dynamics, bedrock feedbacks, and the effect of atmospheric carbon dioxide concentration.

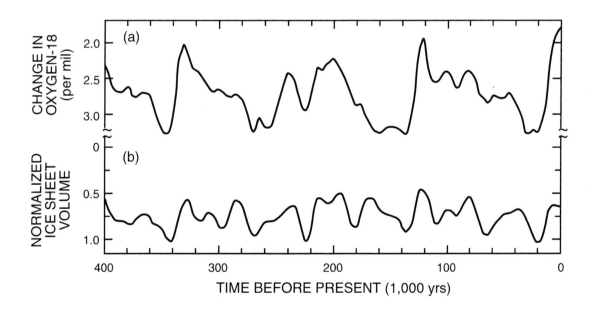

Figure 14. A comparison of (a) oxygen-18 record from Hays et al. (1976) with (b) an ice age simulation of Northern Hemisphere ice volume from the energy balance climate model with an explicit ice sheet formulation from Pollard et al. (1980). Note the correspondence of secondary cycles but the absence of the dominant ramp-like glacial-interglacial signal in the model simulation.

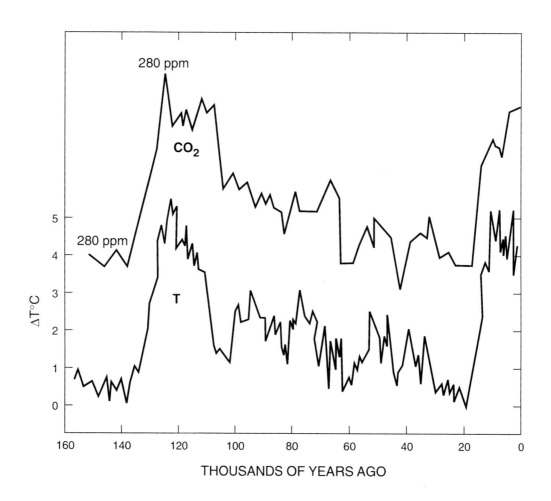

Figure 15. A comparison of the temperature history of Earth over the last 160,000 years with the carbon dioxide history recorded in bubbles trapped in glacial ice.

Summary

The climatic variation preserved in rocks and sediments is the product of complex interactions involving Earth's two major heat engines, the Sun and the radioactivity of Earth's interior. Changes in the amount and distribution of solar energy received at the top of the atmosphere cause changes in climate through their influence on the circulation of the oceans and atmosphere. The energy from Earth's interior fuels plate tectonics, leading to changes in continental positions, volcanism, mountain building, and sea-level variations, which in turn modify climate.

The two case studies we have examined illustrate the importance of these two major heat engines in governing past climates: (1) The dramatically different geography of the Cretaceous resulted in global warming. The increased rate of sea-floor spreading also resulted in higher levels of atmospheric carbon, which promoted additional warming and resulted in one of the warmest climates that can be well documented. (2) During the last 1.9 million years (the Pleistocene) changes in Earth's orbit resulted in variations in the distribution and amount of incoming solar energy, causing rhythmic glacial advances and retreats. The glacial advances are associated with decreases in atmospheric carbon dioxide, thus amplifying the global cooling associated with the changes in orbit. Each case study indicates that several factors are required to explain the climate of any time period, and, significantly, the mechanisms are interrelated. In the Cretaceous, plate tectonics altered climate directly by modifying the oceanic and atmospheric circulation and indirectly by changing the composition of the atmosphere. In the Pleistocene, the changes in Earth's orbit modified the climate directly by changing the incoming solar energy, and indirectly by climate-related changes in the composition of the atmosphere.

Both extensive observations of the past and the application of climate models that incorporate a representation of how the climate system operates are required to decipher the ancient climate record. Knowledge of how the climate of Earth has responded in the past contributes to our understanding of the sensitivity of our climate to change.

GLOSSARY

accretion—the addition of new oceanic crust at the margins of a divergent plate boundary (e.g., midocean ridge) by the cooling of molten material

albedo—the ratio of the amount of radiation reflected by a body to the amount incident upon it

ecotherm—a cold-blooded organism

flux—the rate of transfer of fluid, particles, or energy across a surface

lapse rate—the change in temperature with height in the atmosphere

latent heat—heat released or absorbed by a system through reversible change of state (solid, liquid, vapor)

lithosphere—the outer solid crust of Earth

sensible heat—energy exchanged by a change in temperature

subduction—the destruction of crustal material at a converging plate boundary where one lithospheric plate is forced beneath the other

surface energy balance—the balance of net solar and infrared radiation, sensible heat, and latent heat at the surface

RECOMMENDED READING

Barron, E.J. Ancient climates: Investigations with climate models. *Reports on Progress in Physics 47* (1985), 1563–99.

——. A warm, equable Cretaceous: The nature of the problem. *Earth Science Reviews 19* (1983), 305–38.

——, C.G.A. Harrison, J.L. Sloan, and W.W. Hay. Paleogeography, 180 million years ago to present. *Ecologae Geol. Helv. 74* (1981), 443–70.

—— and W.M. Washington. The role of geographic variables in explaining paleoclimates: Results from Cretaceous climate model sensitivity studies. *J. Geophys. Res. 89* (1984), 1267–79.

Berger, A. Milankovitch theory and climate. *Reviews of Geophysics 26*(4) (1988), 624–57.

COHMAP Members. Climatic changes of the last 18,000 years: Observations and model simulations. *Science 241* (1988), 1043–52.

Crowley, T.J., D.A. Short, J.G. Mengel, and G.R. North. Role of seasonality in the evolution of climate during the last 100 million years. *Science 231* (1986), 579–84.

Emiliani, C. The causes of the ice ages. *Earth and Planetary Science Letters 37* (1978), 349–52.

Hays, J.D., J. Imbrie, and N.J. Shackleton. Variations in the Earth's orbit: Pacemaker of the ice ages. *Science 194* (1976): 1121–32.

Imbrie, J., and K.P. Imbrie. *Ice Ages, Solving the Mystery.* Short Hills, New Jersey: Enslow Publishing, 1979.

Oerlemans, J. Model experiments on the 100,000-year glacial cycle. *Nature 287* (1980), 430–32.

Pollard, D. A simple ice sheet model yields realistic 100 kyr glacial cycles. *Nature 296* (1982), 334–38.

——, A.P. Ingersoll, and J.G. Lockwood. Response of a zonal climate–ice sheet model to the orbital perturbation, during the Quaternary ice ages. *Tellus 32* (1980), 301–19.

INDEX

albedo, 6

carbon dioxide
    Cretaceous, 12
    Plate tectonics, 6, 9
    Pleistocene, 18–20
climate change
    albedo, 6
    carbon dioxide, 9, 12, 18–20
    causes, 6–9, 14–17
    continental distribution, 6–7, 10
    continental elevation, 8–10
    ice-bedrock effects, 18–19
    land area, high latitudes, 6–7, 10
    ocean currents, 7
    one hundred thousand year rhythms, 18–19
    orbital variations, 14–20
    regional, 8, 10
    sea level, 11
    space scales, 1
    thermal inertia, 7, 11
    time scales, 1, 3, 13–20
climate models, 9, 18
    carbon dioxide, 12, 18
    Cretaceous, 10–12
    geography, 10–11
    orbital variations, 18–20
Cretaceous
    climate, 3, 10–12
    geography, 6, 10
    sea level, 5
    temperatures, 3–4, 12
    time scale, 4
Crowley, Thomas, 1 1

elevation, 8–9, 10
energy budget
    surface, 7
    latent heat, 7
    sensible heat, 7

general circulation
    easterlies, 9
    topographic effects, 8
    westerlies, 9
geologic time scale, 4
    Cretaceous, 4
    Pleistocene, 4
glacial cycles, 13–20
    rhythms, 14–20
    theories, 14
glaciation
    glacial record, 4
    ice cap formation, 13
    moraines, 13
    rhythms, 13–14

Hays, James, 14, 17–18
heat transport
    oceans, 7
Held, Isaac, 18

ice caps
    formation, 13
Imbrie, John, 14, 17–18
isotopes
    oxygen, 13–14

lapse rate, 8

Milankovitch, Milvtin, 14, 17–18

ocean currents, 7
Oerlemans, Johannes, 18
orbital elements, 14–17
    climate sensitivity, 18–19
    eccentricity, 15, 18
    obliquity, 16, 18
    precession of the equinoxes, 16, 18
    tilt, 15, 18

plate tectonics, 1, 3
    accretion, 5
    carbon dioxide, 9
    climate change, 6–9, 12
    mid-ocean ridges, 5
    Pangaea, 5
    Panthalassa, 5
    sea floor spreading, 5, 9
    subduction, 5
Pleistocene
    climate, 13–20
    oxygen isotopes, 14
    rhythms, 13–14
    time scale, 4

regional climate
    topographic effects, 8

sea level, 5, 9, 11
Shackleton, Nick, 14, 17–18
solar insolation, 15, 17
Suarez, Max, 18

thermal inertia, 7

warmth
    extremes, 2
    evidence for, 3, 4